Oliver Lodge

On a Dynamo for Maintaining Electrical Vibrations of High Frequency

Elibron Classics
www.elibron.com

Elibron Classics series.

ISBN 0-543-70351-7 (paperback)
ISBN 0-543-70350-9 (hardcover)

This Elibron Classics Replica Edition is an unabridged facsimile of the edition published in 1913 in London.

From the PHILOSOPHICAL MAGAZINE *for June* 1913.

ON A DYNAMO FOR MAINTAINING ELECTRICAL VIBRATIONS OF HIGH FREQUENCY.

WITH SOME NOTES ON THE TRANSMISSION OF WAVES IN WIRELESS TELEGRAPHY.

BY

SIR OLIVER LODGE.

From the PHILOSOPHICAL MAGAZINE *for June* 1913.

On a Dynamo for maintaining Electrical Vibrations of High Trequency. With some notes on the Transmission of Waves in Wireless Telegraphy. By Sir OLIVER LODGE.

I DESIRE not so much to put on record, since that has in some sort been already done, but to welcome the achievement of what many physicists in years past desired to attain, namely the construction of an alternating arrangement capable of maintaining electrical vibrations of sufficiently high frequency to give a continuous train of waves of moderate length; comparable for instance to four miles or even less. It may be that this problem will give way, as problems so often do, in several directions at once; and that the result may be—possibly has been—achieved by several different methods. At present, however, my attention has been called to the ingenious device introduced by Dr. Goldschmidt of Berlin, an ex-Professor of Electrotechnics at Darmstadt.

A method of achieving the desired end by a series of alternating dynamos had already been suggested, and will be found referred to in Professor G. F. FitzGerald's Collected Papers, page 280, where a discussion between himself and Dr. Sumpner, involving also Professor Trouton, is reported as having occurred at the Physical Society of London on 22nd January, 1892; with myself, as it happened, in the chair. It is reported also in *Nature*, vol. xlv. p. 358.

The plan then mooted was to use the alternating current from one dynamo to excite the field magnets of another, that other to excite a third; and so on. By this means it was clearly perceived that the frequency could be made to mount up in arithmetical progression with the number of machines. But the obvious cumbersomeness and wastefulness of the arrangement, although it was true that all the dynamos might run on the same shaft, caused the idea to be promptly abandoned.

Dr. Goldschmidt, however, has now made the interesting discovery that all these steps can be taken in a single machine, with some manifest and some unexpected advantages, and apparently with sufficient economy for practical purposes even at high power, so long as the final frequency does not exceed say 50,000 per second.

The fact that iron is used, and to all appearances necessarily used, in attaining this result, although its magnetization should not be pushed beyond a permeability of 100 or something below the steep part of a magnetization curve, necessitates many instrumental precautions, much careful design, and considerable constructive skill; but designers of the Firm of Messrs. C. Lorenz, and Engineers of the Firm of Messrs. Bergmann, both in Berlin, have been able to execute machines which certainly work in the way expected.

So far as I am aware Physicists in this country have not made themselves generally acquainted with the device employed, and inasmuch as its ingenuity commands admiration, I venture

(1) to expound the principle briefly,
(2) to attempt a theory.

The features of the machine and the laws utilised may be all considered known when taken separately, some of them well known; and yet, in combination, they work in a way which at first sight seems puzzling and might readily have been pronounced fanciful.

The separate points on which the invention is based may be thus stated:—

(1) That when the field magnet of an alternating dynamo, with a number of poles distributed round a circle, is itself excited by an alternating current, its field may be considered as revolving equally in two opposite directions; *i. e.*, its magnetism revolves although the iron is mechanically stationary. This is merely because each North pole changes sinuously into a South pole, and each South pole into a North; thus virtually changing places in a continuous cyclical manner.

(2) The second point is that an armature revolving in front of such a field magnet, which if the magnet were steady would have the frequency q generated in it, will, when the field magnet itself is excited by an alternating current of frequency p, be subject to the resultant frequencies $p-q$ and $p+q$.

For if the armature stood still, the induced frequency would be p; and if it revolves, its own frequency q is superposed upon this; in one direction running after and perhaps overtaking the frequency p, and in the other direction running away from and thereby virtually increasing the relative magnetic velocity.

If the two frequencies are equal, being both generated by the highest rate of rotation mechanically safe, the two resultant frequencies will be 0 and $2p$.

This is the principle which it was proposed to utilise in the series of machines mentioned above as under discussion in 1892.

(3) This is not real doubling, but an addition of one to one, as can be proved by continuing the process and employing the frequency $2p$ as again a field magnet exciter; for, when an armature is revolving in front of that, the frequencies generated in that new rotor are $3p$ and p respectively (the sum and difference). Hence the progression is merely arithmetical, not geometrical. If, still further, the $3p$ is used as exciter, the frequencies appearing in a new rotor will be $4p$ and $2p$; and so on.

(4) The frequency 0 in the first rotor is of course not a current at all—though it sounds like a continuous current—but it is static magnetism, and may be said to correspond with the natural reaction between field magnet and armature, like that between a magnet and its keeper, when both are stationary.

(5) The $2p$ generated in the second armature as the difference between stator and rotor frequencies is likely to be opposite in phase to the $2p$ generated in the same armature as the sum of rotor and stator frequencies; and the two may therefore partially cancel each other. Similarly with other intermediate frequencies in the chain, each of which occurs twice over. The only unique currents will be the first and the last—the steady current in the magnets of the original exciter, and the highest frequency currents in stator and in rotor at the other end of the chain.

(6) The self-induction of the different coils, which would naturally cause lag and entail considerable complication and impedance, can be neutralised for any desired frequency by

the insertion of capacity in series; so that the general equation for a circuit, with $\dot{u}$ as current,

$$L\ddot{u}+R\dot{u}+\frac{u}{S}=E\cos pt$$

becomes simplified, when the current is properly attuned to the driven frequency, *i. e.* tuned in such a way that $p^2\,LS=1$. It then, after a brief moment of growth, assumes the simple form

$$\dot{u}=\frac{E}{R}\cos pt\,;$$

because the first and third terms neutralise each other under tuned conditions; since then $\ddot{u}+p^2u=0$.

(7) If a current is started and stopped in such a circuit, its rise and fall are governed by the customary laws; for it must be remembered that there are free oscillation terms in the complete solution which control the initial and final stages of such a maintained current, and that these are discarded from the permanent theory as being too rapidly evanescent to deserve much notice. The initial (and final) free oscillations are rapidly killed out of the complete solution of the above equation by an exponential time-coefficient involving R/2L, and forced oscillations alone survive. But the free are not wholly to be ignored, and in some cases may be found to give trouble. The current starting from zero, when it is attuned to the maintained oscillations and not over-damped, rises to its final stage according to the equation

$$\dot{u}=\frac{E}{R}(1-e^{\frac{Rt}{2L}})\cos pt,$$

as clearly explained in Lord Rayleigh's 'Sound,' vol. i. §§ 47, 48. Hence, to attain rapid response, it may even be desirable to introduce a little extra resistance into some of the circuits, so as to diminish the time-constant 2L/R by which these momentary perturbations are regulated. It may be noted, as a practical consequence of the completer theory, that the frequency factors should be chosen with capacity large and inductance small; for it is the latter factor alone that retards the attainment of the permanent state.

Of course a current takes time to develop in a circuit, even without any self-induction,—the velocity of light must be involved and a travelling wave-front; but in ordinary practice these can safely be ignored.

(8) The current and the E.M.F will be in the same phase, and lag will be obliterated, if the precise frequency is maintained.

(9) The system is not self-exciting but requires an initial starting current or an excited magnet, without which the whole is inactive. The initial current supplies no energy however, and, if no variations were wanted, its place might be taken by a permanent magnet. All the effective energy of the system is derived from the mechanical rotation; and signalling can be managed by varying the exciting current.

(10) The original exciter may be either a steady current or a comparatively slowly alternating one, say of acoustic frequency. It might for instance be supplied through a microphone and be carved into fluctuations by speech, which could thus apparently be effectively magnified and indirectly transmitted to a distance. The frequency of the exciting stimulus is of no consequence so long as it is incomparably slower than any of the really high frequencies to which the circuits are tuned.

An exciting current of acoustic frequency, say 500 a second, is practically a steady stimulus of slowly varying strength during each half period; but the result of employing such a current as the stimulus may be the powerful emission of a note of 1000 per second, when the perfectly inaudible electric waves are collected and reinterpreted by some device.

(11) The neutralisation of self-induction by capacity applies not only to the whole of a circuit but to any part of it for which the condition corresponding to $p^2\mathrm{LS}=1$ is satisfied; and the only obstacle in that part of a circuit is then its ohmic resistance, subject of course to any throttling or skin effect due to the most favourable automatic distribution of current in the substance of the conductor; a distribution which can be made virtually uniform by adequate subdivision, as by employing a cable composed of fine insulated wires. In so far as iron is used, the resistance term will contain eddy-current losses and hysteresis, which must be guarded against in known ways.

(12) That if to a circuit with one L and S in series, another L and S pair is added also in series, the natural frequency is unaltered; for though the resulting inductance is doubled, the resulting capacity is halved, and the product remains the same. The advantage of such a doubly-tuned circuit is that its parts are connected by nodes, to which anything can be attached without disturbance; for a pair of

nodes might even be short-circuited without altering the frequency of vibration.

(13) A tuned or resonant circuit can be maintained in vibration with very little power, the only expenditure of energy being that concerned with the dissipation term RC^2, together with a similar expression for the energy radiated away which in the case of an open circuit may become important.

In some cases wave radiation is the thing desired, and in that case the expenditure in this way is useful dissipation, and the greater its proportional amount the higher the efficiency of the arrangement.

(14) That in an alternating dynamo, consisting of a rotor and a stator, so far as mutual action is concerned it does not matter which is which ; we are only concerned with the relative motion between the two, and each can react on the other by mutual induction.

Putting all these simple principles together, and providing both the rotor and the stator with as many tuned branch circuits as are required, Dr. Goldschmidt has designed his dynamo, whereby by mutual action and reaction between stator and rotor the frequency mounts to a level such that unavoidable hysteresis and other iron losses make further additions to it uneconomical or impracticable. And the highest frequency can be used for emission of waves.

In a machine which now exists at Slough, the revolutions being about 4000 a minute, the frequency attained directly by multiple poles and high speed rotation is 12,000 per second ; and this by action and reaction is multiplied by 4 ; thus emitting a wave from the aerial $6\frac{1}{4}$ kilometres or about 4 miles long.

A skeleton diagram of Dr. Goldschmidt's ingenious arrangement can be given thus: without any statement that those precise arrangements are necessarily adhered to. The rotor connexions are supposed to be made with slip rings as usual: the whole rotor, and whole stator, being of course connected up, and not a portion only as in the diagram.

An explanation is hardly necessary, since each branch circuit is labelled with the frequency-constant to which it is attuned, and nodes are plainly provided for the attachment of capacity in series so as to raise the frequency in some branches.

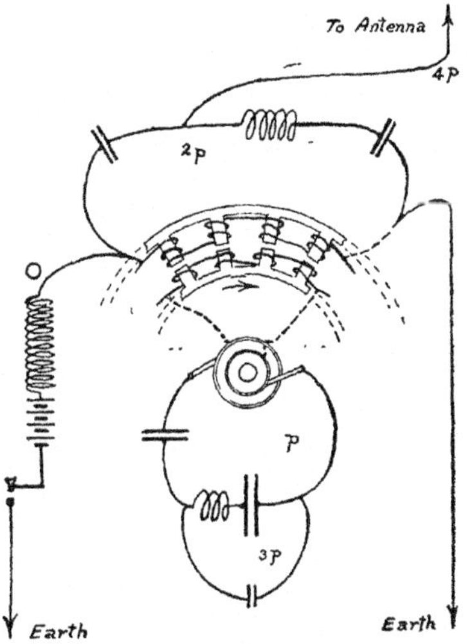

Diagram to illustrate principles used in the Goldschmidt Dynamo, not any actual arrangements.

Theory of the Goldschmidt Dynamo.

The theory as verbally explained in the preceding pages appears to be expressible by the two following equations, subject always to satisfaction of tuning conditions in every circuit so that the self-inductance and capacities neutralise each other throughout.

$$x = \frac{M}{R}\frac{d}{dt}(y \cos pt) \quad . \quad . \quad . \quad . \quad . \quad . \quad (1)$$

$$y = y_0 + \frac{M}{R'}\frac{d}{dt}(x \cos pt), \quad . \quad . \quad . \quad . \quad . \quad (2)$$

where x is the current in the rotor, y in the stator, and M is the amplitude of the mutual induction between them, *i. e.* the value when the poles are opposite each other. At other times the mutual induction is M cos pt.

And since the currents are variable too, the above equations are justified. They look simple enough, but to solve them generally is not easy; though it appears physically necessary that

x must develop into a series of circular functions of odd multiples of pt, while

y develops into a similar series with even multiples of the same quantity.

The series may be stopped at any stage by ceasing to provide resonating circuits and permitting radiation and dissipation of energy.

To prove the above equations—remembering that they only have that simple form when tuned circuits are provided for every intended frequency—write the number of lines of force from stator to rotor as N; and from rotor to stator as N′; then the E.M.F. generated is dN/dt or dN'/dt respectively.

But $N = my$, and $N' = mx$,

and $m = M \cos pt$, *i. e.* is a sinuous function of time as the armature revolves.

Hence, if the rotor circuit resistance is R, while that of the stator is R′, and if the conditions of § (6) are satisfied,

$$x = \frac{E}{R} = \frac{1}{R}\frac{dN}{dt} = \frac{1}{R}\frac{d}{dt}(my) = \frac{M}{R}\frac{d}{dt}(y \cos pt).$$

Similarly, if y_0 is the stimulating current initially supplied to the stator, the current in it is

$$y = y_0 + \frac{M}{R'}\frac{d}{dt}(x \cos pt).$$

So they are proved. I have not yet succeeded in solving them generally, but it is not difficult to proceed by stages.

Working out of the Equations.

To deal with these equations easily, take them piecemeal, and apply them to say four provided branch circuits beside the original exciting circuit of stator. Two of these tuned circuits are connected to rotor, with resistances R_1 and R_3, two with the stator, with resistances R_2 and R_4. The latter circuit involves useful dissipation by radiation as well as all hysteresis and ohmic losses.

The initial current in stator being y_0, the current first generated in rotor is

$$x_1 = \frac{1}{R}\frac{dN}{dt} = -\frac{pM}{R_1} y_0 \sin pt.$$

The current generated by reaction from this, back into the stator, is

$$y_2=\frac{M}{R_2}\frac{d}{dt}(x_1\cos pt)$$

$$=\frac{p^2M^2}{R_1R_2}y_0\sin^2 pt-\frac{p^2M^2}{R_1R_2}y_0\cos^2 pt$$

$$=-\frac{p^2M^2}{R_1R_2}y_0\cos 2pt.$$

The current which this now generates in the rotor is

$$x_3=\frac{M}{R_3}\frac{d}{dt}(y_2\cos pt)$$

$$=\frac{p^3M^3}{R_1R_2R_3}y_0\cos 2pt\sin pt+\frac{2p^3M^3}{R_1R_2R_3}y_0\sin 2pt\cos pt$$

$$=\frac{p^3M^3}{R_1R_2R_3}y_0\{\sin 3pt+\sin 2pt\cos pt\}$$

$$=\frac{p^3M^3}{R_1R_2R_3}y_0(1\tfrac{1}{2}\sin 3pt+\tfrac{1}{2}\sin pt).$$

The current which x_3 now generates in the stator is

$$y_4=\frac{M}{R_4}\frac{d}{dt}(x_3\cos pt)$$

$$=\frac{M}{R_4}\cos pt\frac{dx_3}{dt}-\frac{pM}{R_4}x_3\sin pt$$

$$=\frac{p^4M^4}{R_1R_2R_3R_4}y_0\left\{\begin{matrix}4\tfrac{1}{2}\cos 3pt\cos pt+\tfrac{1}{2}\cos^2 pt-\sin 3pt\sin pt\\ -\sin 2pt\cos pt\sin pt\end{matrix}\right\}$$

$$=A\{4\tfrac{1}{2}\cos 3pt\cos pt-\sin 3pt\sin pt+\tfrac{1}{2}\cos^2 pt-\tfrac{1}{2}\sin^2 2pt\}.$$

Employing the identity $\cos^2 A-\sin^2 B=\cos(A+B)\cos(A-B)$ this becomes

$$y_4=A(4\cos 3pt\cos pt+\cos 3pt\cos pt-\sin 3pt\sin pt)$$

$$=A(4\cos 3pt\cos pt+\cos 4pt)$$

$$=A\{2(\cos 4pt+\cos 2pt)+\cos 4pt\}$$

$$=\frac{p^4M^4}{R_1R_2R_3R_4}y_0(3\cos 4pt+2\cos 2pt).$$

Hence the *total* current in rotor is

$$x=x_1+x_3=y_0\left\{\frac{p^3M^3}{R_1R_2R_3}\cdot\frac{3}{2}\sin 3pt+\frac{pM}{R_1}\left(\frac{p^2M^2}{2R_2R_3}-1\right)\sin pt\right\}; . \quad (3)$$

and the *total* current in stator is

$$y=y_0+y_2+y_4=y_0\left\{\frac{p^4M^4}{R_1R_2R_3R_4}\,3\cos 4pt+\frac{p^2M^2}{R_1R_2}\left(\frac{2p^2M^2}{R_3R_4}-1\right)\cos 2pt+1\right\} \quad (4)$$

Thus y_0, the original exciting current, is a simple factor throughout; and the circular functions of even and odd multiples of pt, which on general grounds were expected, are obvious.

A rather more general treatment.

On showing this to my brother, Alfred Lodge, of Charterhouse, late of Coopers Hill, he pointed out that the neatest substitution would be

$$\theta=pt+\tfrac{1}{2}\pi,$$

so as to throw everything into the cosine form and get a general method of formation; subsequent re-interpretation being as follows:—

$$\cos\theta=-\sin pt;\quad \cos 3\theta=\sin 3pt;$$

$$\cos 2\theta=-\cos pt;\quad \cos 4\theta=\cos 4pt.$$

The terms in the pt formulæ will all be of the same sign, all the x series being in sines, and all the y series in cosines of multiples of pt; while the θ formulæ are all in cosines, the signs of successive terms being alternately + and −.

If all the coefficients affected with the + sign are added together, their sum will always exceed the sum of the − coefficients by unity; a fact which furnishes a check on arithmetical correctness.

To calculate successive x's and y's proceed thus:—

Write

$$\theta=pt+\tfrac{1}{2}\pi.$$

Then

$$x_1=\frac{M}{R_1}\frac{d}{dt}(y_0\cos pt)=\frac{pM}{R_1}\frac{d}{d\theta}(y_0\sin\theta)$$

$$=c_1y_0\cos\theta.$$

$$y_2=\frac{pM}{R_2}\frac{d}{d\theta}(\sin\theta\,.\,x_1)=c_1c_2y_0\frac{d}{d\theta}(\tfrac{1}{2}\sin 2\theta)$$

$$=c_1c_2y_0\cos 2\theta.$$

$$x_3 = \frac{pM}{R_3}\frac{d}{d\theta}(\sin\theta\,.\,y_1) = c_1c_2c_3y_0\frac{d}{d\theta}(\sin\theta\,.\cos 2\theta)$$

$$= c_1c_2c_3y_0\,.\,\frac{1}{2}\frac{d}{d\theta}(\sin 3\theta - \sin\theta)$$

$$= \tfrac{1}{2}c_1c_2c_3y_0(3\cos 3\theta - \cos\theta).$$

Let us ignore $c_1c_2c_3\ldots y_0$ for the present. They evidently come in as multiples only.

$$y_1 = \frac{1}{2}\frac{d}{d\theta}(\sin\theta\,.\,3\cos 3\theta - \sin\theta\cos\theta)$$

$$= \frac{1}{4}\frac{d}{d\theta}(3\sin 4\theta - 4\sin 2\theta)$$

$$= \frac{1}{4}(3\times 4\cos 4\theta - 4\times 2\cos 2\theta)$$

$$= 3\cos 4\theta - 2\cos 2\theta.$$

Now suppose this is $A_4\cos 4\theta - B_4\cos 2\theta$.

$$x_5 = \frac{d}{d\theta}\sin\theta(A_4\cos 4\theta - B_4\cos 2\theta)$$

$$= \frac{1}{2}\frac{d}{d\theta}\{A_4\sin 5\theta - (A_4 + B_4)\sin 3\theta + B_4\sin\theta\}$$

$$= \frac{1}{2}\{5A_4\cos 5\theta - 3(A_4 + B_4)\cos 3\theta + B_4\cos\theta\}$$

$$= \frac{1}{2}(15\cos 5\theta - 15\cos 3\theta + 2\cos\theta).$$

Call this $A_5\cos 5\theta - B_5\cos 3\theta + C_5\cos\theta$, and note that this makes $A_5 - B_5 + C_5 = A_4 - B_4 = 1$.

Similarly

$$y_6 = \tfrac{1}{2}\{6A_5\cos 6\theta - 4(A_5 + B_5)\cos 4\theta + 2(B_5 + C_5)\cos 2\theta\}$$

$$= \quad 3A_5\cos 6\theta - 2(A_5 + B_5)\cos 4\theta + (B_5 + C_5)\cos 2\theta$$

$$= \tfrac{1}{2}(45\cos 6\theta - 60\cos 4\theta + 17\cos 2\theta).$$

Also

$$x_7 = \tfrac{1}{4}\{45\times 7\cos 7\theta - 105\times 5\cos 5\theta + 77\times 3\cos 3\theta - 17\cos\theta\}$$

$$= \tfrac{1}{4}\{315\cos 7\theta - 525\cos 5\theta + 231\cos 3\theta - 17\cos\theta\},$$

and

$$y_8 = \tfrac{1}{8}\{315\times 8\cos 8\theta - 840\times 6\cos 6\theta + 756\times 4\cos 4\theta - 248\times 2\cos 2\theta\}$$

$$= \quad 315\cos 8\theta - 630\cos 6\theta + 378\cos 4\theta - 62\cos 2\theta,$$

with multiplier $c_1c_2c_3c_4c_5c_6c_7c_8\,y_0$.

Note that 315+378 exceeds 630+62 by unity; also that

$$A_8 = 1\times\tfrac{3}{2}\times2\times\tfrac{5}{2}\times3\times\tfrac{7}{2}\times4 = \underline{|8} \div 2^7.$$

Continuing, we find that the positive coefficients for x_9 are the halves of 2835, 5040, and 62; while the negative ones are the halves of 6615 and 1320.

For y_{10} the positive coefficients are the halves of 14175, 34965, and 1382; while the negative ones are the halves of 37800 and 12720.

The law of formation of the terms is fairly simple:—
Thus let

$$u_{n-1} = A_{n-1}\cos(n-1)\theta - B_{n-1}\cos(n-3)\theta + C_{n-1}\cos(n-5)\theta - + \ldots\ldots,$$

then

$$u_n = \tfrac{1}{2}\{A_{n-1}\,.\,n\cos n\theta - (A_{n-1}+B_{n-1})(n-2)\cos(n-2)\theta + \\ = A_n\cos n\theta - B_n\cos(n-2)\theta + - \ldots\ldots$$

multiplied in each case by $y_0\,c_0\,c_1\,c_2\ldots\ldots$; the last factor being c_n in the case of y_n. Also $A_n = n\,! \div 2^{n-1}$.

A few minor considerations applicable in a 4 circuit machine.

In first writing down the general equations (1) and (2) for the arrangement diagrammatically illustrated above, the fact that the different branch circuits may have different resistances was ignored. No principle is thus neglected, and general solution is made easier. The only resistance which is not unimportant or unessential from the point of view of radiation is the last of the series, R_4, which ought to be large compared with the others.

The constant which characterizes every expression is of the form pM/R, the frequency-constant multiplied by the mutual time-constant of each circuit. There are 4 circuit branches altogether, besides the steady exciting-current circuit of the stator; the circuits of the rotor have been called R_1 and R_3, those of the stator have been called R_2 and R_4; the latter circuit involves useful dissipation by radiation as well as all hysteresis and ohmic losses.

Thus to summarise sufficiently, ignoring differences in R for the moment, call $pM/R=c$, for all four circuits.

$$x=y_0\{\tfrac{3}{2}c^3 \sin 3pt+c(\tfrac{1}{2}c^2-1) \sin pt\} \quad . \quad . \quad . \quad (3')$$

$$y=y_0\{3c^4 \cos 4pt+c^2(2c^2-1) \cos 2pt+1\}. \quad . \quad (4')$$

There are really four characteristic constants

$$\frac{pM}{R_1}, \quad \frac{pM}{R_2}, \quad \frac{pM}{R_3}, \quad \frac{pM}{R_4},$$

which we may denote by c_1, c_2, c_3, c_4, respectively, and so can write the values

$$x=c_1y_0\{\tfrac{3}{2}c_2c_3 \sin 3pt+(\tfrac{1}{2}c_2c_3-1) \sin pt\} \quad . \quad . \quad (3'')$$

$$y-y_0=c_1c_2y_0\{3c_3c_4 \cos 4pt+(2c_3c_4-1) \cos 2pt\}. \quad . \quad (4'')$$

To cancel the p frequency, c_2c_3 must equal 2, or $p^2M^2=2R_2R_3$.

To cancel the $2p$ frequency, c_3c_4 must equal $\frac{1}{2}$, or $p^2M^2=\frac{1}{2}R_3R_4$.

To cancel both, $2R_2R_3$ must equal $\frac{1}{2}R_3R_4$, or $R_4=4R_2$.

It is not likely that these cancelling conditions can be satisfied; nor is it at all certain that it would be desirable to satisfy them, since possibly if any intermediate step were omitted the graduated rise of frequency might cease. But the partial cancelling of merely auxiliary currents is of interest, and ought to have some effect in reducing unnecessary iron losses. The highest frequency losses are inevitable, and the limit of practicable frequency will be reached when hysteresis and other such losses become too prominent in comparison with what is available for radiation. But it must always be remembered that the shorter the wave the more powerful is the emission; and so, presumably, the shorter the wave the more vigorous would be the reception at a distance, save for difficulties connected with transmission and accidents by the way.

Part II. On the Transmission of Waves in Air.

The rate of emission of energy from an aerial was calculated by FitzGerald for a magnetic radiator in 1883, and by Hertz for an electric one in his great paper of 1888. Hertz's result is that the loss of energy per half swing is

$$\frac{8\pi^4 e^2 l^2}{3K\lambda^3}$$

and the loss per second therefore

$$\left(\frac{2\pi}{\lambda}\right)^4 \cdot (el)^2 \cdot \frac{v}{3K} = \text{energy radiated per second}$$

by an electric radiator of moment el. It is easy to get this by integrating the flow of energy per unit area, $EH/4\pi$, all over the surface of a large sphere.

For a magnetic or closed circuit radiator, of moment μnAC, a similar expression can be written with v/μ instead of v/K.

This may be thrown into various convenient forms (see Lodge and Howard, Phil. Mag. July 1889), and the form I prefer in practice is in terms of the mean square of current supplied to the aerial:

$$\begin{array}{c}\text{Radiation Power}\\ \text{in kilowatts}\end{array} = 4\,(\text{amperes})^2 \left(\frac{\text{height}}{\text{wave-length}}\right)^2 \quad . . \ (5)$$

because the current in an aerial can be measured with an ammeter—if necessary a sensitive hot-wire instrument heavily shunted,—and the radiation power can thus be calculated. Another way of measuring the current would involve the use of a flowing-water calorimeter; which I propose to try for this purpose.

The main uncertainty about the data is an estimate of the effective height of the upper capacity area which shall properly enter as a factor into the aerial's electric moment el; unless indeed there is also a lower capacity area thoroughly insulated from the ground: for the effective height must partly depend on the perfection or imperfection of the earth-connexion and the quality of the soil. If the earth were a perfect conductor the effective height would be twice the apparent height. In practice there is generally some intermediate value characteristic of the circumstances of the station.

To establish the above very handy expression for the energy actually radiated per second, proceed thus:—

The magnetic field close to a linear Hertz oscillator is

$$H = \frac{eln}{r^2} \cos nt \sin \theta;$$

which corresponds to an oscillating current of strength $en \cos nt$ and length l.

The current amplitude is therefore

$$C_0 = en = \frac{2\pi e}{T} = \frac{2\pi ev}{\lambda}.$$

The average current depends on how it is measured; its square may be $\frac{1}{2}C_0^2$. But if it is considered as the quantity passing in half a swing, $C = e/\frac{1}{2}T$, the radiation expression becomes

$$\frac{16\pi^4 e^2 l^2 v}{3K\lambda^4} = \frac{4\pi^4 C^2 l^2}{3Kv\lambda^2} = \frac{400\,\mu v C^2 l^2}{3\lambda^2} = \frac{400 \times 10^{10} \mu C^2 l^2}{\lambda^2} \text{ ergs per sec.}$$

But in electromagnetic measure the unit of current is 10 amperes, and 10^{10} ergs per second is a kilowatt, so the formula for radiation power is proved; and the true radiation (subject to correction for the numerical factor) is

$$4 \times (\text{mean amperes})^2 \times \frac{l^2}{\lambda^2} \text{ kilowatts}, \quad . \quad . \quad . \quad (5)$$

l being the effective height, or linear factor of the electric moment, of the radiator.

There are many circumstances which it would seem necessary to take into account besides those which appear in this formula. Capacity, for instance, and self-induction, and arrangement of antenna generally. But they are all implicitly involved in the wave-length, and there is no need specially to attend to them except in considering the design of an antenna. For that purpose it must be remembered that in true waves the electric and magnetic energies are equal, and in any region where they are not equal the excess of one is useless for wave emission. At great distances the electric and magnetic intensities are necessarily and automatically equal, because any other part of the energy has returned to the oscillator and is promoting persistence of vibration, which in the case of shock-excitation is a useful thing to do.

But to determine the conditions which assist radiation we may calculate separately the electric and magnetic intensities at points on the equator fairly near the oscillator, within say a quarter wave-length: or perhaps most simply at the critical distance $r = \lambda/\pi\sqrt{2}$, characterized by $m^2r^2 = 2$, the place at which progressive radiation really begins.

It is manifest that the shorter the wave-length the stronger will these intensities be, and hence the stronger the radiation, other things being equal; for the radiation through unit area at every place is by Poynting's theorem $EH/4\pi$.

Consider therefore the circumstances on the equator of a

Hertz oscillator in general. The electric field is there all axial, and is given by

$$E=\frac{el}{Kr^3}\left\{(1-m^2r^2)\sin(mr-nt)-mr\cos(mr-nt)\right\},$$

which may be written with a phase angle $(mr-\phi)$ and a clearly expressed amplitude, thus:—

$$E=\frac{el}{Kr^3}\sqrt{(1-m^2r^2+m^4r^4)}\sin\{nt-(mr-\phi)\},$$

where $\tan\phi=\frac{mr}{1-m^2r^2}$.

The magnetic intensity at the same place lies along the equator and is

$$H=\frac{elmn}{r}\left\{\sin(mr-nt)+\frac{\cos(mr-nt)}{mr}\right\};$$

or, rewriting it,

$$H=\frac{eln}{r^2}\sqrt{(1+m^2r^2)}\sin\{nt-(mr+\phi')\},$$

where $\tan\phi'=\frac{1}{mr}$.

So at a distance $r=\lambda/\pi\sqrt{2}$, or $\lambda/4{\cdot}4$, the amplitude of the electric intensity is

$$E_0=\frac{el}{Kr^3}\sqrt{(1-2+4)}=\frac{el\sqrt{3}}{Kr^3},$$

while the magnetic amplitude is

$$H_0=\frac{eln}{r^2}\sqrt{(1+2)}=\frac{enl\sqrt{3}}{r^2}.$$

Expressing these in terms of current amplitude, whose value may be written alternatively

$$C_0=en=\frac{2\pi e}{T}=\frac{2\pi ev}{\lambda}=emv,$$

we get, at a distance such that $mr=\sqrt{2}$ or $r=\lambda/\pi\sqrt{2}$,

$$H_0=\frac{\sqrt{3}\,.\,C_0l}{r^2}=\frac{\sqrt{12}\,.\,\pi^2lC_0}{\lambda^2},$$

$$E_0=\frac{\sqrt{3}\,.\,l}{Kr^3}\cdot\frac{C_0}{mv}=\frac{\sqrt{3}\,.\,l\,.\,m^2C_0}{2^{\frac{3}{2}}Kv}=\frac{\sqrt{6}\,.\,\pi^2l\mu vC_0}{\lambda^2}.$$

To express C_0 in magnetic units we take μ as 1, and to express E_0 in electrostatic units we divide by v.

Consequently, in the units most convenient,

$$E_0=\sqrt{\frac{3}{2}}\cdot\frac{C_0 l}{r^2}$$

while $$H_0=\sqrt{3}\cdot\frac{C_0 l}{r^2}.$$

Thus for a simple vertical linear oscillator, the magnetic force somewhat exceeds the electric in the region where waves begin. Hence anything which tends to increase the electric force, such as a tilted capacity area or the like, will assist radiation in one general direction:—as initiated by Dr. de Forest in 1904, introduced by Mr. Marconi in 1905, and described by him to the Royal Society on 22 March, 1906 (Proc. Roy. Soc.).

Design of Aerial.

The necessity for superposed electric and magnetic fields, and the fact that radiating power is their vector product, at right angles to the plane of both, explains the main behaviour of directive antennæ. A linear Hertz vibrator, for instance, gives no axial radiation for two reasons, (1) because there is no magnetic field at the axis, and (2) because the electric force there is "end-on," so to speak, and could not contribute to axial radiation anyhow. So also in the case of a closed-circuit oscillator, say a large plane coil of one turn; though the magnetic force along the axis is strong, its direction prevents its contributing to axial radiation. Hence if a linear and a closed-circuit oscillator are combined, as they practically are in a Γ shaped aerial, especially when the upper arm is very long, the fore and aft radiation in the plane of the Γ, notably that in front of the vertical part, will be copious compared with that in a perpendicular direction; and the interesting figure-of-8 equal-intensity region, detected by Mr. Marconi and his staff, is a consequence.

All increase of capacity combined with height, in the aerial, goes to increase both the electric and the magnetic fields at a distance, and therefore will be advantageous (though expensive) notwithstanding the concurrent increase of wave-length. In fact free capacity and height afford two separate means of increasing wave-length without diminishing the power of emitting radiation; while both together increase that power; for radiation is proportional to the square of SVl/λ^2, and λ^2 is proportional to S and also to l if the radiator is straight.

But insertion of inductance coils in the aerial will have

the effect of lengthening the wave without strengthening the field at a fixed distance, and so must act deleteriously on radiation intensity; though such coils will promote persistent oscillation and tuning, and may therefore be advantageous up to a point. Their use will always be of the nature of a compromise; and this whether they are actually inserted in the aerial or are inductively associated with it: so they should be kept to a reasonable minimum.

In an umbrella aerial whose ribs approach the ground, charge from the region of greatest capacity has to ascend before descending; wherefore it fails to exert full magnetic force at a distance, and conserves some energy which it might radiate, thereby becoming more capable of tuning. The amount of potential tuning necessary or desirable in an antenna must depend on the method of excitation. With shock-excitation some amount of it is essential; with maintained waves it appears to be unnecessary; and accordingly in the latter case the more closely an antenna can approach to a simple Hertz oscillator—a copious radiator and absorber—the better; some kind of directivity being a still further improvement when circumstances allow.

A practical remark may here be interpolated:—The importance of great height for the upper capacity area is so striking, and the trouble and expense of securing it by artificial structures so obvious, that I doubt if sufficient attention has been paid—in the selection of a long-distance site—to the possibility of finding a mountain or rock of sufficiently non-conducting material to serve as the supporter of the upper part of an antenna for a station at its base. Experience would be necessary to ascertain whether the surgings in such a hill, or pair of twin hills, need be any more deleterious to efficient radiation and reception than they are in the iron of the Eiffel Tower, for instance.

Magnetic Radiation.

A magnetic radiator only differs from an electric radiator in having as moment a current area instead of a pair of separated charges; *i. e.* μAC instead of el.

So if the value of C is made to correspond with en, which would usually mean a very strong current since n is very great, the two radiators would be equally effective when AC in magnetic measure equals el in electric; in other words when

$$2\pi A = l\lambda,$$

i. e. in the case of a Γ aerial, when the efficient length of upper arm equals $\lambda/2\pi$.

To see what $C=en$ means. Write $e=SV$.

Let S be comparable to Kl', that is let it correspond to a capacity area not greatly differing from the height of the aerial in linear dimensions, and let V correspond say to nearly an inch spark, say 60,000 volts or 200 electrostatic units. Then

$$\begin{aligned} C &= nSV \\ &= \frac{2\pi}{T} \cdot \frac{l'}{v^2} \cdot 200\, v \text{ magnetic units} \\ &= \frac{4000\,\pi\, l'}{\lambda} \text{ amperes.} \end{aligned}$$

If l' is 200 metres and $\lambda = 4$ kilometres, $C=600$ amperes.

It is noteworthy, though now familiar, how strong a current amplitude is readily produced by a disruptive discharge.

Arguments for use of long waves.

Although short waves have so many advantages, they need not be as efficient as long ones at really great distances if they have to overcome obstructions or opacities in the atmosphere. For when it comes to transmitting waves near the surface of the earth, the simple theory of unimpeded etherial transmission is no longer sufficient. Earth conduction and air conduction become of the essence of the phenomenon, and many complications may happen to the waves, some of which have attracted the attention of mathematicians—notably at present Professor Zenneck, who has dealt elaborately with the problem of earth and sea conduction in his important paper in the *Annalen der Physik*, vol. xxiii. page 846, 1907; summarised or translated by Professor Fleming in 'Engineering' for June 4th and 11th, 1909, and also in the 1910 edition of his book. His treatment of the effect of earth-surface conductivity seems very thorough, but as regards the effect of aerial conductivity our knowledge is still slight.

The reason of the observed obstructive influence of sunshine is not yet clearly known; but if the opacity of ordinary air due to observed ionization is calculated from recognized data, it appears to be by no means sufficient to account for the observed effect by simple obstruction: though the surprisingly high conductivity of rarefied air in the absence of electrodes must not be forgotten.

On the whole it seems probable that the difficulty of transmission by day must be due to something happening in the upper conducting layer of the atmosphere. I assume that grazing reflexion from this layer must usually assist transmission. If the transition of density is sudden, the layer will constitute a better reflector than if it were gradual; and if it were not sudden enough, either greater power or longer wave-length would help to overcome the difficulty. If, however, under the influence of rising sunlight or heat the reflecting layer should become puckered, so as to offer something nearer normal incidence to approaching waves,—thereby letting them escape into space,—no increase of power alone will do very much towards remedying the evil; and increased wave-length may become a necessity.

To examine this whole question further:—

An absorbed plane wave equation is

$$\frac{d^2F}{dr^2}=\mu K\frac{d^2F}{dt^2}+\frac{4\pi\mu}{\sigma}\frac{dF}{dt};$$

and for a solution there are two extreme cases according to whether a critical number

$$\frac{4\pi}{\sigma p K} \quad \text{or} \quad \frac{4\pi v^2\mu T}{2\pi\sigma} \quad \text{or} \quad \frac{2v\lambda}{\sigma/\mu}$$

is small or large. (See for instance Lodge "On Opacity" in Phil. Mag. for April, 1899.)

Now the ionization of sun-lighted air under some conditions has been estimated as 10^{-12} of the total number of molecules present: if so it will have a specific resistivity

$$\sigma=2\times10^{20}\ \mu \text{ sq. centimetres per second.}$$

In that case the critical number will be small, being of the order 10^{-5} for a wave comparable with a kilometre in length.

The form of solution applicable in that case is

$$F=F_0 e^{-\frac{2\pi\mu v}{\sigma}r}\cos(mr-pt).$$

So the distance required to reduce the amplitude of these true waves to $1/e$th would then be independent of the frequency, *i. e.* would be the same for waves of any length, and its value is

$$r_1=\frac{\sigma}{2\pi\mu v}.$$

Inserting the above estimate of σ for ionized air, this comes out

$$r_1=\frac{1{\cdot}8\times10^{20}}{18\times10^{10}}=10^9 \text{ centimetres}$$

$$=10{,}000 \text{ kilometres.}$$

The ionization and conductivity of air encountered by the waves must therefore manifestly be greater than that—*i. e.* the resistivity must be smaller than 10^{20} c.g.s.—if the atmosphere is to stop waves by sheer opacity.

But for a conductivity like that of sea-water (which may well be reached and considerably exceeded in the atmosphere, even without special ionization, by mere reduction of density, at the best conducting horizontal level—at least when the electric force surpasses a certain critical intensity below which even rarefied air must insulate unless ionized—),

say $\sigma = 3 \times 10^{10}$ μ sq. centimetres per second,

the critical number,

$$\frac{2v\lambda}{\sigma/\mu} = \frac{2\lambda}{1 \text{ centimetre}},$$

is certainly big; and in that case the other or diffusion solution applies, namely

$$F = e^{-\sqrt{\left(\frac{2\pi\mu p}{\sigma}\right)}r} \cos\left\{pt - \sqrt{\left(\frac{2\pi\mu p}{\sigma}\right)}r\right\};$$

and the distance for reduction of amplitude to $1/e$th of its value would then depend on the frequency, for it is

$$r_1 = \sqrt{\left(\frac{\sigma}{2\pi\mu p}\right)} = \frac{1}{2\pi}\sqrt{\left(\frac{\sigma\lambda}{\mu v}\right)} = \sqrt{\left(\lambda \times \frac{1}{40} \text{centim.}\right)}.$$

For a 4 kilometre wave this distance is 1 metre; and the absorption constant varies inversely with the square root of the wave-length. Differently sized waves, moreover, would travel at different rates; and accordingly the trouble of *distortion,* so familiar in cables, would be introduced by high atmospheric conductivity. But we have no knowledge of the actual existence of such conductivity except at great elevation, where, if it formed a sharply defined smooth horizontal layer it would be a help, not a hindrance, and would assist in overcoming the curvature of the earth—either for the mirage-like reason suggested by Dr. Eccles or by still more simple reflexion.

A law of absorption dependent on the inverse root of the wave-length has been suggested by observation; and such a law indicates greater effective conductivity in the general atmosphere than is usually considered likely.

Connexion between Ionization and Conductivity.

It may be convenient here to write out the connexion between ionization and conductivity for air, as ordinarily treated electrolytically.

Let n be the number of ions of one sign per c.c. supposed equal, e the charge on each, and u $(=u+v)$ their combined velocities.

The intensity of current, $\frac{\text{current}}{\text{area}} = i = neu.$

The dissociation ratio n/N we may call k, N being the

total molecules per c.c.; and by Professor Townsend's measurements the value of Ne for air at ordinary pressure and temperature is $1{\cdot}22\times10^{10}$ E.S. units per c.c., or $\cdot41$ E.M. units. So, since $e=4{\cdot}65\times10^{-10}$ E.S. units, $N=2{\cdot}6\times10^{19}$.

The speed of ions in air at ordinary pressure, when urged by a force X, has been measured, and is given by

$$u_1=\frac{u+v}{X}=420+510=930 \text{ centim. per sec. per unit E.S.F.,}$$

$$=3{\cdot}1 \text{ centim. per second for a gradient of 1 volt per centim.}$$

It is found that over a considerable range of pressure u_1 varies inversely with the concentration, or Nu_1 is constant. So, from the above data, $Neu_1=4{\cdot}1\times3{\cdot}1=13$ amperes per square centimetre when driven by a volt per centimetre. The intensity of current $i=kNeuX$, so the specific conductivity

$$\frac{i}{X}=\frac{1}{\sigma}=kNeu,$$

and $$\frac{1}{k\sigma}=13\,\frac{\text{amperes}}{\text{sq. centimetre}}\div\frac{1\text{ volt}}{\text{centimetre}}.$$

Hence $$\sigma=\frac{1}{13k}\text{ ohm centimetre, or }\frac{10^9}{13k}\text{ c.g.s.}$$

and the number of ions per c.c. is

$$n=kN=\frac{2{\cdot}6\times10^{19}\times10^9}{13\sigma}=\frac{2}{\sigma}\times10^{27}$$

when σ is expressed in absolute c.g.s. units.

At really low pressures, comparable with a millimetre of mercury or less, the mobility of ions is much greater than what is here dealt with. Besides, all this refers to atomic or molecular ions, *i.e.* to electrolytic conduction, and does not take into account the far greater mobility of free electrons, even though they be only momentarily or transitionally or potentially free. It seems possible that some writers, in considering the opacity of air, may have been satisfied with the known kinds of ion for which data were available; and that the presence of free electrons in the upper regions would give a much higher conductivity than has been suspected. It can hardly be doubtful that electrons are directly supplied to the upper air, in the form of β rays from the sun.

Ionization Causes.

With reference to the likely amount of ionization in the upper atmosphere, I doubt if sufficient attention has been paid to the positive charges presumably received from the sun. If the sun is radioactive it probably emits α as well as

β rays; and though the β rays, being most penetrating, seem to have been most studied, the α particles may be really more important. The β rays will tend to be magnetically deflected towards the poles, presumably returning along upper strata; the α particles will travel with the beam of light and will be stopped in the upper regions of the atmosphere, where they will tend to accumulate. If any γ rays exist in solar radiation,—and if β rays escape from the solar atmosphere surely γ rays will—they will exert a deep-seated effect on our atmosphere; and a horizontal beam of them might produce a fairly vertical stratum of ionized air, such as would constitute a fairly effective screen or obstacle to waves which are dependent on reflexion from an upper region for their advance.

Living under a reflecting dome it is natural that there should be occasional local concentration of signals, and that the varying shape and quality of the dome should greatly influence the signals locally received. Earth or sea conductivity has its more uniform effect, and assuredly conveys some part of the signals; but sea-water conduction is not likely to be the cause of capriciousness in signalling, nor would it contribute to daylight effects, nor to those specifically observed at sunrise and sunset. All those must be due to atmosphere.

As regards practical telegraphy—on which here I am only touching very incidentally—it by no means follows that earth conduction and earth tapping ought not to be more employed for signalling purposes than they are; perhaps already they are giving more assistance to long-distance signalling than is recognized. It is quite possible that, by the aid of electric vibrations maintained in the earth's crust, aerial or Hertzian waves may ultimately be dispensed with, and something less liable to stray or varying atmospheric influences be employed. The trouble of *distortion*, to which this method would be liable, could be overcome by emitting only a single pure wave. Meanwhile, and so long as Hertzian waves are used, it is safe to say that if by reason of inequalities at sunrise the best conducting layer of the atmosphere becomes corrugated or bent, it is quite likely to divert a good deal of radiant energy which at sufficiently oblique incidence would be reflected: and it would be most deadly to the shorter waves. Also, the effect of a vertical conducting layer, or of corrugations in a horizontal one, would be likely to be felt most when they were moderately near a sending or a receiving station, and would probably be less noticeable when the abnormalities were diffused over the half-way region; though an observation of Mr. Marconi in a Royal Institution lecture to the effect that the signals were weakest between Clifden and Glace Bay when the boundary between day and night was half-way across the Atlantic, tends to negative and reverse this idea.

If anything like a vertical reflector existed behind a sending station (as just before sunrise, for instance) it might lead to a temporary intensification of distant response. Furthermore, any kind of opacity which increased with the intensity of radiation,—and we shall proceed in a sequel to consider such kinds,—would be more effective near a sending than near a receiving station. The fact that sunrise or sunset at either station—*i.e.* a horizontal beam—is specially effective, whether as assistance or as obstruction, and the maxima and minima which then occur, are very suggestive. For facts of observation, see first the original discovery by Mr. Marconi of the effect of sunshine, in the course of his fine experiment of sending Hertz waves across the Atlantic and detecting them at different distances and under different conditions all the way across; as described by him in Proc. Roy. Soc. June 1902; also his Nobel Prize lecture, December 1909, quoted by Fleming, 'Wireless Telegraphy,' p. 826 (1910 edition). On the same subject Mr. Pickard made observations in America, which are quoted by Professor Pierce in his book on Wireless Telegraphy, page 135; Dr. Austin, of the U.S. Naval Wireless Department, made important measurements in 1910 which are quoted in Erskine Murray's book, page 298 of the 1913 edition. In this book indeed there is a general discussion of the whole subject, with a summary of practical experience; and there are doubtless other data now available. Some recent ones by Messrs. Round and Tremellen are recorded in the 'Marconigraph' of November 1912, where also a British Association discussion at Dundee is reported. Special attention must be called to the Addresses of Prof. Fleming and Dr. Eccles reported in *Nature*, xc. p. 262, and in *Electrician*, 27 Sept. 1912; also to Dr. Eccles' paper in Proc. Roy. Soc. June 1912.

I see that Professor Zenneck in his important paper in the *Annalen der Physik*, vol. xxiii. p. 862, makes a very low estimate of the conductivity of ordinary air, and doubts if it can effectively obstruct the passage of waves. But the fact which he notes, that mist-laden air is more transparent to long-distance signalling than clear air, speaks powerfully in favour of the hostile influence of ionic conduction, or at least of the adverse influence of free electric charges; since in mist the ions would have moisture condensed round them, and would therefore be loaded so as to destroy the ready mobility on which conducting power, or any influence which they could exert, must depend. Thus whether ions act by ordinary conductivity, or in some other way such as is considered below, we can at least say that opacity and conductivity are similarly affected, both being reduced by the presence of mist.

Going back to the opacity equation, pp. 776, 7, and its two cases there considered, there is one intermediate possibility; namely that the resistivity of the medium might sometimes have a value such as would make the critical number neither large nor small, but say unity. We may consider what will happen then.

If $4\pi/\sigma p K$ is to be 1, it follows that $\sigma = 2\mu v\lambda$; or, for a kilometre wave, that σ is of the order

$$6 \times 10^{15}\ \mu \text{ sq. centimetres per second.}$$

This appears a very possible value for rarefied air;—indeed Dr. Schuster has estimated that, at a height of 100 kilometres, atmospheric conductivity is of the order 10^{-13} c.g.s., equivalent to 10,000 ohms per c.c., in order to satisfy the requirements of his theory of magnetic variations;—and in that case we must use the general solution

$$F = F_0 e^{-\alpha r} \cos(pt - \beta r)$$

with $$\alpha^2 \text{ or } \beta^2 = \frac{1}{2}\mu K p^2 \left\{ \sqrt{\left(1 + \left(\frac{4\pi}{\sigma p K}\right)^2\right)} \mp 1 \right\}.$$

Or calling the critical constant c,

$$\alpha = \frac{\pi}{\lambda}\left\{2\sqrt{(1+c^2)} - 2\right\}^{\frac{1}{2}}$$

while $$\beta = \frac{\pi}{\lambda}\left\{2\sqrt{(1+c^2)} + 2\right\}^{\frac{1}{2}}.$$

These are general values; so, taking c unity, as the special case we are now considering, we get

$$\alpha = \sqrt{(2\sqrt{2} - 2)}\frac{\pi}{\lambda} \doteq \cdot 9\frac{\pi}{\lambda},$$

$$\beta = \sqrt{(2\sqrt{2} + 2)}\frac{\pi}{\lambda} \doteq 2{\cdot}1\frac{\pi}{\lambda}.$$

Hence, with the intermediate value for σ, the distance for reduction of amplitude to $1/e$th is comparable with the wave-length; being, in the special case just written, about $\lambda/3$. The absorption constant begins in this case to be inversely as the root of the wave-length, and there is incipient departure from true wave propagation.

Any local approach to such atmospheric conductivity as belongs to the intermediate case,—for instance by the stirring up of some highly ionized layer by convection currents,—would amply account for serious obstruction to any but the longest waves.

Other Causes of Atmospheric Opacity.

Apart from what is caused by regular electric conductivity in the ordinary way, there is another possible kind of opacity, due to a curious and unexpected effect of radiation on electrons, discovered mathematically as far as I know by Sir J. J. Thomson in 1902, and contained in his paper "On the Emission of Negatively Electrified Corpuscles by Hot Bodies," in the Philosophical Magazine for August 1902 (ser. 6, vol. iv. p. 258). This paper appears to have attracted very little attention, though it is referred to by Professor Fleming in his comprehensive treatise on Wireless Telegraphy, 1910 edition, page 824.

The foundation of it is the action of etherial waves in propelling or sweeping up electrified particles in the direction of wave propagation, thereby communicating to them energy which presumably must be taken from the available energy of the waves; though the author of the theory does not seem to attach importance to this process as an absorber of energy, nor to contemplate it from that point of view. He directs attention to another and more ordinary kind of opacity, due to the scattering of radiation by corpuscles over which the waves are passing. So far as *light* is concerned, this process is indeed likely to be more effective than the other as a disperser of radiant energy, though it leaves the energy in the ether and does not transfer it to particles of matter. The radiation scattered by the particles should be polarized, as all light reflected from small particles is, and Sir J. J. Thomson calculates the coefficient of absorption due to this kind of scattering, as

$$\frac{4\pi\mu^2 N e^4}{3m^2}.$$

This therefore is independent of wave-length, but must always be very small, even when the ions are electrons, because the fourth power of ionic charge occurs in the numerator. The ionization might be great, say a million electrons to the cubic centimetre, and yet the above absorption-coefficient would be only of the order 10^{-20} per centimetre. This is insignificant of course for all terrestrial distances, but in the depths of space it seems likely to have an effect of some importance, not indeed as destroying energy in the ether, but as ultimately converting transparency into translucency.

What the ionization of interstellar space is I have no

idea, but assuming there may be no more than four stray electrons on the average to the cubic centimetre, there would still be a calculable opacity due to this cause; for a distance of 10^{21} kilometres, or a hundred million light-years, would reduce the energy of direct radiation to $1/e$th of its value.

Professor Barkla has informed me that for X-ray pulses it is not necessary for electrons to be isolated in order to exert a scattering influence, but that electrons associated with atoms of matter appear to be sufficiently free to take part in the process. It may be that thus much ultra-violet light is stopped, for to small enough waves the opacity so caused would be considerable, because of the immense value of N. That which is supposed by the above theory to occur to waves of all kinds, in really ionized space, has thus been already observed in the case of those minute waves to which atom-bound electrons are partially and sufficiently free.

Returning to the consideration of the other phenomenon, the bodily propulsion or sweeping forward of electrons by the wave: at first sight it is not easy to realize physically why electrified particles should be thus propelled; but no flaw appears in the reasoning, and it may be classified as a phenomenon akin to the magnetic deflexion of cathode rays. For (*a*) The electric force of a wave will obviously tend to move a charged particle in its own direction in the wave-front; (*b*) directly it moves, it is a current, and is therefore acted on by the magnetic component of the wave; and (*c*) this action will propel it in a direction at right angles to both, *i. e.* normal to the wave-front.

Taking the plane of the wave as xy, and the electric intensity as X, the force acting on a charge e is Xe in the first instance; though in so far as the particle is propelled along the ray it becomes subject to magnetic force also, and the whole x force is

$$m\ddot{x} = \mathrm{X}e - \mu \mathrm{H}e\dot{z}.$$

Also
$$m\ddot{y} = 0,$$

and
$$m\ddot{z} = \mu \mathrm{H}e\dot{x}.$$

These are the equations; and when they are worked out, as they are on pages 257–8 of the Philosophical Magazine for August 1902, the result is a progression superposed upon an oscillation of twice the wave-frequency, a sort of fore-shortened cycloidal motion, expressed by the equation:—

$$v^2 - \left(v - \frac{dz}{dt}\right)^2 = \frac{e^2\mu^2\mathrm{H}_0{}^2\lambda^2}{8\pi^2 m^2}\left\{1 - \cos 2(pt - qz)\right\},$$

the original wave being

$$\mathrm{H} = \mathrm{H}_0 \cos (pt - qz).$$

Hence, denoting the quantity $\frac{e\mu H_0\lambda}{m\pi\sqrt{8}}$ by u, and the maximum or ultimate value of $\frac{dz}{dt}$, apart from oscillations, by w, we have the equation

$$v^2-(v-w)^2=u^2;$$

or, since u is generally decidedly smaller than v,

$$w=\frac{u^2}{2v}=\frac{e^2\mu^2H_0^2\lambda^2}{16\pi^2m^2v}.$$

The particle will be joggled forward with twice this speed every time a ventral segment reaches it, and will be reduced to momentary rest in the suck-back of the nodes.

On showing this to Professor Poynting he stated his belief that a resistance term had been omitted from the original equations, a term expressive of the energy consumed in radiation whenever a charge is accelerated. Indeed a moving charge seems likely to be in a curiously unstable condition, for although it encounters no resistance as long as its velocity is uniform, yet directly there is acceleration of any kind, even retardation, it must begin to transfer its energy to the ether in a way which is equivalent to the action of a retarding force; so that an initial retardation would tend to be intensified, and thus apparently an electron could be readily induced to stop dead. If this be the case, then the opacity hereafter treated may depend not merely on the newly born ions, as hereafter supposed, but on the total ionic, or at least electronic, content: and the sweeping up of the ions would itself constantly make a call upon the energy of the propelling waves, unless their motion could be guaranteed absolutely uniform. I do not propose to attempt to develop this important suggestion, and will proceed without further reference to it.

The inverse electrochemical equivalent e/m is $10^7/\sqrt{\mu}$ c.g.s. units, and the wave-energy per c.c. is $\mu H_0^2/8\pi$; (remembering the electric energy as well as the magnetic, and also the fact that H_0 is the maximum, not the average, of H).

So $$w=\frac{\mu e^2}{m^2}\cdot\frac{\mu H_0^2}{8\pi}\cdot\frac{\lambda^2}{2\pi v};$$

and the velocity imparted to the particles is proportional both to the energy-density and to the square of the wave-length. Short waves, like those of light, cannot be expected to do much in the way of propulsion, except perhaps in the neighbourhood of the sun where they are very intense; but anything like radiation of Hertzian wave-length may be expected to sweep away the charged corpuscles which it encounters and to leave a clear space.

To get an idea of the magnitude of this influence of a Hertz vibrator, take a straight vertical one of electric moment Ql with a current amplitude in it $C_0=2\pi Q/T$. The magnetic field at any distance r (greater than say half a wave-length) is

$$H=\frac{Qlpq}{r}\cos\theta\,.\,\sin(qr-pt),$$

where θ is the inclination to the horizon (the latitude, so to speak). So on the "equator" the magnetic amplitude is

$$H_0=\frac{4\pi^2 Ql}{T\lambda r}=\frac{2\pi C_0 l}{\lambda r}.$$

Hence the velocity acquired by the particles,

$$w=\left(\frac{\mu}{4\pi}\cdot\frac{e}{m}\right)^2\frac{(\lambda H_0)^2}{v}$$

$$=\frac{\mu^2}{4v}\left(\frac{e}{m}\right)^2\left(\frac{C_0 l}{r}\right)^2.$$

So the speed generated in an electron by an upright Hertz radiator, at a distance 10 times the height of the aerial along its equator, will be comparable to a kilometre per second if the mean aerial current is 600 amperes. The velocity thus generated naturally falls off with the square of the distance, just as the wave-energy does.

The mere propulsion of a set of particles will not appreciably consume the energy of a wave, provided the particles remain individually the same and continue free, for electrons steadily on the run would not contribute to opacity*. Under the influence of Hertzian waves they would either be driven into the atoms, or else would speedily acquire a terminal velocity and be swept out of the way. The process suggests a curious method of clearing the air of electric charges; and there may be some application of long electric waves for this purpose. Assuming that ions in the atmosphere must have some meteorological influence, it would seem just possible that an indirect ether-wave effect on the weather, such as cranks have anticipated as likely to be due to wireless telegraphy, may not be so absurd as at first sight it appeared.

But passing this for the moment, and returning to the

* See, however, the suggestion of Professor Poynting, referred to on page 784.

question of a possible opacity or obstruction to the passage of waves due to this cause, we must realize that solar radiation consists not of light alone but probably of α, β, and γ rays also. Ultra-violet light, and α rays if they reach us, will be active ionizers in the upper regions of air, but the more penetrating kinds of radiation should ionize much more than a superficial layer, and when the beam of sunshine is horizontal a great deal of atmosphere may be affected.

By some such extra-terrestrial influence, in addition to any effect of emanations from the earth's crust, the ionization of the atmosphere may be preserved and constantly renewed, in spite of the clearing-away tendency of Hertzian waves. And if these waves thus have a continually renewed set of charged particles to deal with, they will be virtually encountering a constantly increasing mass; and some amount of their momentum and energy must be absorbed by transference to the particles.

Now we know, from Maxwell, Poynting, and others, that the energy of radiation per unit volume may be regarded as a momentum carried across unit area per second; in other words, as a pressure or potential pressure in the direction of travel. This pressure P may be expressed either as an energy density, *i. e.* in ergs per c.c.; or as a pressure, in dynes per square centimetre; or as a surface density multiplied by an acceleration; or as a volume density multiplied by the square of a velocity; or merely as the momentum travelling per second through unit area. If no obstacle is encountered, the radiation momentum is merely passed on in the ether undiminished, and the potential pressure is not mechanically exerted; but if matter is present, some of it is transferred to matter, and the etherial power will to that extent diminish, even without any production of heat.

It has long been known that small particles are most readily moved by the pressure of light, but now we see that charged ions of low atomic weight, and especially electrons, are more affected than anything else; especially by long waves.

The momentum transmitted per second to matter will be a certain fraction of $\mathrm{P}v$, and will equal the material momentum generated in the same time, $\mathrm{N}mvw$; where mw is the momentum attained by each particle, and $\mathrm{N}v$ the number freshly encountered per unit area per second.

We may express it better thus:—

The loss of P per centimetre is $d\mathrm{P}/dx$, and this will be accounted for by the gain of momentum generated per second in each cubic centimetre, $\frac{d}{dt}(\mathrm{N}mw)$. Ordinarily, in such an expression, speed is the variable; but here w is a terminal velocity, which must ultimately be attained by all the fresh matter encountered, so N is the variable; and $d\mathrm{N}/dt$ signifies the number of ions continually produced by extraneous causes in each c.c. per second. Hence our equation is

$$\frac{d\mathrm{P}}{dz} = -mw\frac{d\mathrm{N}}{dt}$$

$$= -\dot{\mathrm{N}}m\,.\,\frac{\mu e^2}{m^2}\,.\,\frac{\lambda^2}{2\pi v}\,.\,\mathrm{P} = \frac{1}{z_0}\mathrm{P}.$$

Or the wave energy-density remaining, after a journey of length z, is

$$\mathrm{P} = \mathrm{P}_0\, e^{-\frac{z}{z_0}}.$$

It must here be remembered that the relevant variation of P has nothing to do with its natural decrease with distance as a function of r, but represents solely its gradual diminution by absorption in successive layers, in geometrical progression. Hence the distance needed to reduce the wave energy per c.c., or the momentum per second per square centimetre, to $1/e$th of the value it would otherwise have had at any given place, is z_0, where

$$z_0 = \frac{2\pi v\, m}{\dot{\mathrm{N}}\,\mu e^2\lambda^2}; \quad . \quad . \quad . \quad . \quad . \quad . \quad (6)$$

which is of *length* dimension, as it ought to be. We see that it will be shorter for electrons than for any other kind of ions, and that the absorption will vary with the square of the wave-length.

To get an idea of its value we can remember

that $\mu e^2/m^2$ is of the order 10^{14} c.g.s., for electrons,
that $v = 3\times10^{10}$ centimetres per second,
that m is of the order 10^{-27} grammes,

and that in practice λ may be anything from 100 metres to 10 kilometres.

Taking these two extreme cases as typical; and assuming, as a moderate estimate, that 10,000 electrons are liberated by sunshine or other agency, per c.c. per second, all along the effective path of the wave; the distance required for etherial-wave-energy to fall to $1/e$th of the value it would otherwise have had is $z_0 = \frac{6\pi\times10^{23}}{\dot{\mathrm{N}}\lambda^2}$ centimetres, or $20\times10^{19}/\lambda^2$.

This distance, therefore, for a 100 metre wave is very great, and for a 10 kilometre wave is 2000 kilometres.

The case of interstellar space.

As to an infinitesimal opacity of interstellar space due to this cause, assuming that any ionization goes on there at all, it is only a question of how long it will be before even light-waves are absorbed; though long waves are much more readily dealt with. Let there be only half an electron produced per c.c. per second, and take the wave-length as 6×10^{-5} centimetre, the absorption-distance in such a case is 10^{33} centimetres, which corresponds to a light journey of 10^{15} years.

The other or scattering kind of absorption above spoken of is therefore likely to be of much greater proportional influence, so far as the short waves of light are concerned; and in the immediate neighbourhood of the sun Sir J. J. Thomson thinks it likely that it may partially account for the Corona.

But in any solar system region of space, where electrons are plentifully produced, long Hertzian waves will be liable to become absorbed and to exhibit their energy in the form merely of streams of corpuscles.

Summary.

Part I. of this paper gives the theory of the Goldschmidt Dynamo, for any number of circuits, up to a magnifying ratio of 10.

Part II. deals with obstacles to electric wave transmission due to various kinds of opacity, or virtual opacity, in the atmosphere; and emphasises the important influence which the highly conducting layer necessarily existing at a certain height must play, with fluctuations according to whether it is uniformly spherical or irregular in shape, and with reference also to the distribution and amount of its ionization. It is suggested that sunshine interferes with this layer; either by altering its shape through heat, or by making its transition more gradual through general supply of β particles. It is also suggested that under the action of solar radiation of all kinds electrons may be momentarily set free enough to promote several newly-expected kinds of atmospheric opacity. Incidentally a useful practical expression for radiation from an aerial is given in equation (5).

Appendix.

More general solution of equations (1) *and* (2) (p. 763).

It was pointed out by Mr. H. F. Thompson of Charterhouse, to whom the equations were shown, that when the coefficient $m=M/R$ was the same in both equations, the variables could be separated by introducing $\xi=x+y$, $\eta=x-y$; for then

$$m\frac{d\xi}{dt}-\xi(\sec pt+mp\tan pt)+b\sec pt=0,$$

$$m\frac{d\eta}{dt}-\eta(\sec pt-mp\tan pt)+b\sec pt=0;$$

whence ultimately

$$\xi\cos\theta=\left(\frac{1+\sin\theta}{1-\sin\theta}\right)^{\frac{1}{2}n}\left\{C-bn\int\left(\frac{1-\sin\theta}{1+\sin\theta}\right)^{\frac{1}{2}n}d\theta\right\},$$

where $\theta=pt$, $n=1/mp$, and $b=y_0$;

with a similar expression for η with the fractions inverted, and a different arbitrary constant. A Fourier expansion can be made, and coefficients calculated; and though there is no guarantee that the series will be convergent, this appears not to matter practically in this case.

It will be more convenient, however, to quote the not dissimilar but completer solution with which Mr. C. A. Gaul of Marlborough has favoured me, for the case when the two coefficients are different. His process, abbreviated, runs closely as follows:—

The given equations are

$$x=m\frac{d}{dt}(y\cos pt);\quad y=b+m'\frac{d}{dt}(x\cos pt);$$

or

$$nx=\cos\theta\frac{d}{ds}(y\cos\theta)\quad.\quad.\quad.\quad.\quad.\quad.\quad.\quad\text{(i.)}$$

and

$$n'y=n'b+\cos\theta\frac{d}{ds}(x\cos\theta),\quad.\quad.\quad.\quad.\quad\text{(ii.)}$$

where $n=1/mp$, $n'=1/m'p$, $\theta=pt$, $s=\sin\theta$.

Now since $\frac{d}{ds}(\cos\theta)=-\tan\theta$, we get from (ii.)

$$n'y\cos\theta=n'b\cos\theta+\cos^3\theta\frac{dx}{ds}-x\sin\theta\cos\theta;$$

substituting in (i.), we can soon find that

$$(1-s^2)^2\frac{d^2x}{ds^2}-4s(1-s^2)\frac{dx}{ds}-x(1+nn'-2s^2)-n'bs=0.$$

In the same way the differential equation for y can be obtained; it only differs in having $+nn'b$ substituted for $-n'bs$.

Therefore x and y are both functions of s.

Hence we may assume that they can be expanded in powers of $\sin\theta$; an expansion which readily takes the form

$$x=a_0+a_1\sin\theta+a_2\cos 2\theta+a_3\sin 3\theta+\ldots\ldots \quad \text{(iii.)}$$

$$y=b_0+b_1\sin\theta+b_2\cos 2\theta+b_3\sin 3\theta+\ldots\ldots \quad \text{(iv.)}$$

Substituting in the original equations, (i.) and (ii.), we get two equations like

$$n(a_0+a_1\sin\theta+a_2\cos 2\theta+a_3\sin 3\theta+\ldots.)$$
$$\equiv-(b_0+\tfrac{1}{2}b_2)\sin\theta+(b_1+b_3)\cos 2\theta-\tfrac{3}{2}(b_2+b_4)\sin 3\theta+\ldots.$$

Equating coefficients in this pair of identities, we find

$$a_0=0, \quad (1) \qquad b_0=b, \quad (2)$$

$$na_1=-(b_0+\tfrac{1}{2}b_2), \quad (3) \qquad n'b_1=-(a_0+\tfrac{1}{2}a_2), \quad (4)$$

$$na_2=b_1+b_3, \quad (5) \qquad n'b_2=a_1+a_3, \quad (6)$$

$$\ldots\ldots\ldots\ldots \qquad \ldots\ldots\ldots\ldots$$

$$na_{2r}=r(b_{2r-1}+b_{2r+1}), \qquad n'b_{2r}=r(a_{2r-1}+a_{2r+1}),$$

$$na_{2r+1}=-\tfrac{1}{2}(2r+1)(b_{2r}+b_{2r+2}), \quad n'b_{2r+1}=-\tfrac{1}{2}(2r+1)(a_{2r}+a_{2r+2}).$$

Taking a_1, b_1, as the arbitrary constants, and using these equations in numerical order, we find in succession b_2, a_2, b_3, a_3 .... in terms of a_1, b_1 and the known constants n, n', b. This is the General Solution.

With $b_1=0$ as the special case, we find successively all the even a's and odd b's zero.

Hence (iii.) and (iv.) reduce to

$$x=a_1\sin\theta+a_3\sin 3\theta+a_5\sin 5\theta+\ldots\ldots \quad \text{(iii'.)}$$

$$y=b_0+b_2\cos 2\theta+b_4\cos 4\theta+\ldots\ldots \quad \text{(iv'.)}$$

where

$$na_1=-(b_0+\tfrac{1}{2}b_2), \qquad b_0=b,$$

$$na_3=-\tfrac{3}{2}(b_2+b_4), \qquad n'b_2=a_1+a_3,$$

$$na_5=-\tfrac{5}{2}(b_4+b_6), \qquad n'b_4=2(a_3+a_5),$$

$$\ldots\ldots\ldots\ldots \qquad \ldots\ldots\ldots\ldots$$

$$na_{2r+1}=-\tfrac{1}{2}(2r+1)(b_{2r}+b_{2r+2}), \qquad n'b_{2r}=r(a_{2r-1}+a_{2r+1}).$$

This solution is excellent, given the legitimacy of a Fourier expansion; but my brother distrusts this as manifestly divergent, and considers that, apart from physical considerations and electrical devices, the strict solution is an exponential one. His own treatment is as follows:—

Given $$nx = \frac{d}{d\theta}(y\cos\theta) \quad . \quad . \quad . \quad . \quad . \quad . \quad . \quad . \quad (1)$$

$$n'(y-b) = \frac{d}{d\theta}(x\cos\theta), \quad . \quad . \quad . \quad . \quad . \quad . \quad . \quad . \quad (2)$$

$$\therefore \quad nn'(y-b) = \frac{d}{d\theta}\left\{\cos\theta\frac{d}{d\theta}(y\cos\theta)\right\}. \quad . \quad . \quad (3)$$

Let $y\cos\theta = z$, $\tan\theta = \sinh u$; and call $nn' = \mathrm{N}^2$.

Whence $\sec\theta = \cosh u = du/d\theta$;

also $$\frac{dz}{du} = \cos\theta\frac{d}{d\theta}(y\cos\theta); \text{ and } \frac{d}{d\theta}\cdot\frac{dz}{du} = \sec\theta\frac{d^2z}{du^2}.$$

Therefore the differential equation becomes

$$\frac{d^2z}{du^2} = \mathrm{N}^2(y-b)\cos\theta; \quad . \quad . \quad . \quad . \quad . \quad . \quad . \quad (4)$$

i. e., $$\frac{d^2z}{du^2} - \mathrm{N}^2z = -\mathrm{N}^2b \operatorname{sech} u = -\mathrm{N}^2b\cdot\frac{2e^{-u}}{1+e^{-2u}}$$

$$= -2\mathrm{N}^2b(e^{-u} - e^{-3u} + e^{-5u} + \ldots).$$

$$\therefore \quad z = \mathrm{A}\cosh \mathrm{N}u + \mathrm{B}\sinh \mathrm{N}u + \frac{2\mathrm{N}^2b}{\mathrm{N}^2 - d^2/du^2}\{e^{-u} - e^{-3u} + e^{-5u} - + \ldots\}. \quad (5)$$

Now $$\frac{1}{\mathrm{N}^2-\mathrm{D}^2}e^{ku} = \frac{e^{ku}}{\mathrm{N}^2-k^2};$$

$$\therefore \quad y\cos\theta = z = \mathrm{A}\cosh \mathrm{N}u + \mathrm{B}\sinh \mathrm{N}u + 2\mathrm{N}^2b\left\{\frac{e^{-u}}{\mathrm{N}^2-1^2} - \frac{e^{-3u}}{\mathrm{N}^2-3^2} + \frac{e^{-5u}}{\mathrm{N}^2-5^2}\cdots\right\}$$

and $$nx\cos\theta = \frac{dz}{du} = \mathrm{N}(\mathrm{A}\sinh \mathrm{N}u + \mathrm{B}\cosh \mathrm{N}u) - 2\mathrm{N}^2b\left\{\frac{e^{-u}}{\mathrm{N}^2-1^2} - \frac{3e^{-3u}}{\mathrm{N}^2-3^2}\cdots\right\},$$

where $\cosh u = \sec\theta = \sec pt$, and $\mathrm{N}^2 = \mathrm{RR}'/p^2\mathrm{M}^2$, as stated above. It can be seen that $e^{-u} = \sqrt{(1-\sin\theta)}/\sqrt{(1+\sin\theta)}$.

That this solution satisfies the equations (1) and (2) can be verified; but the series of circular functions is more practically useful, since the higher frequencies which go to build up the above total result are barred by the conditions of the actual problem.

Printed in Dunstable, United Kingdom

91219473R00030